THE FARMINGTON COMMUNITY LIBRARY
FARMINGTON HILLS BRANCH
32737 West Twelve Mile Road
Farmington Hills, MI 48334-3302

FEB 2 1 2007

Reading Essentials in Science

GLOBAL ISSUES

Global Warming

Karen E. Bledsoe

PERFECTION LEARNING®

Editorial Director: Susan C. Thies
Editor: Paula J. Reece
Design Director: Randy Messer
Book Design: Emily J. Greazel
Cover Design: Michael A. Aspengren

A special thanks to the following for his scientific review of the book:
Wayne B. Merkley, Professor of Biology, Drake University

Image Credits:
© ARGOSY: p. 6; © CORBIS: p. 10; © Roger Ressmeyer/CORBIS: pp. 14, 35; © Ted Spiegel/CORBIS: p. 15; © Bettmann/CORBIS: p. 19; © Yogi, Inc./CORBIS: p. 21; © AFP/CORBIS: pp. 34, 37

ArtToday (some images copyright www.arttoday.com): back cover, pp. 8, 28, 39 (background), 39 (inset), 41 (top left, background, center), 43; Dynamic Graphics: p. 20–21, 41 (top right); CORBIS Royalty Free: cover (main, bottom right), pp. 1, 11, 33 (top left), 33 (top right), 38, 46–47, 48; Corel Professional Photos: front cover (bottom left, bottom center), pp. 2–3, 5, 7, 20, 26, 27, 30, 31, 42, 44–45; Perfection Learning Corporation: pp. 9, 16, 17, 29, 33 (bottom), 40; MapArt: p. 25

Text © 2004 by Perfection Learning® Corporation.
All rights reserved. No part of this book may be reproduced, stored in a retrieval system, or transmitted in any form or by any means, electronic, mechanical, photocopying, recording, or otherwise, without prior permission of the publisher.
Printed in the United States of America.

For information, contact
Perfection Learning® Corporation
1000 North Second Avenue, P.O. Box 500
Logan, Iowa 51546-0500.
Phone: 1-800-831-4190
Fax: 1-800-543-2745
perfectionlearning.com

1 2 3 4 5 6 BA 08 07 06 05 04 03
ISBN 0-7891-6035-8

Table of Contents

1. The Changing World 4
2. What Causes the Greenhouse Effect? 8
3. Carbon Dioxide and Fossil Fuels 13
4. The Changing Climate 18
5. The Effects of Changing Temperatures 23
6. What Does the Future Hold? 32
   Conclusion 43
   Internet Connections and Related Readings . . . 44
   Glossary 46
   Index 48

Chapter One

The Changing World

In 1988, countries all around the world experienced unusual weather. The number of forest fires rose. **Drought** made the fires worse. Some countries were hit hard with floods and **hurricanes**. It seemed as though the weather had gone crazy.

In the same year, a scientist at NASA spoke before the U.S. Congress. He said that he was "99 percent confident" that the entire globe was warming up. Higher global temperatures were changing the weather.

Other scientists agree. While they disagree about what the causes are and the effects will be, most agree that the Earth is warming up.

What's it stand for?

NASA stands for National Aeronautics and Space Administration. It was created in 1958 in the United States.

How much has the Earth warmed?

Over the past 100 years, the Earth has warmed about 1°F. Doesn't sound like much? Think about this. Since the last **ice age**, 18,000 to 20,000 years ago, the Earth has only warmed 5 to 9 degrees.

Going back 600 years, studies show that the 20th century was the warmest century. Not only that, but 1990, 1995, and 1997 were the warmest years in that 600-year span. Some researchers claim that there is a 90 percent chance that the Earth's average temperature will rise by 3 to 9 degrees over the next 100 years!

Any change in the average global temperature causes changes in the **climate**. The climate of a region is the long-term average weather conditions. Changes in climate, such as rainfall, temperature, and wind patterns, affect all living things. But how is the Earth warming? Scientists have named the culprit the ***greenhouse effect***.

What is a greenhouse?

A greenhouse is a small building enclosed in glass used for growing plants.

When was the greenhouse effect discovered?

To hear people talk, you may think that the greenhouse effect is a new discovery. But scientists have known about it for almost two centuries.

5

In the early 1800s, French mathematician Joseph Fourier was studying heat flow in the **atmosphere**. The atmosphere is a mixture of gases that surrounds and protects the Earth. Fourier discovered that the atmosphere can trap heat just as the glass sides of a greenhouse can. Heat passes through the glass of a greenhouse and warms the inside. Some of the heat escapes back through the glass, but some is trapped inside. The air inside the greenhouse becomes warmer than the air outside. Fourier discovered that the Earth is warmed in the same way.

Just what is CO_2?

Carbon dioxide is often written as CO_2. The C means that carbon dioxide contains one atom of carbon, and the O_2 means that it contains two atoms of oxygen.

In 1896, Swedish chemist Svante Arrhenius described how Fourier's discovery worked. Arrhenius said that **carbon dioxide** and several other gases are **greenhouse gases**.

Carbon dioxide, Arrhenius said, is given off by living things as they breathe out. The carbon comes from materials living things consume. Carbon is absorbed by the ocean. It is also taken in by plants and algae. This cycling of carbon, Arrhenius said, keeps the level of carbon dioxide in the atmosphere in

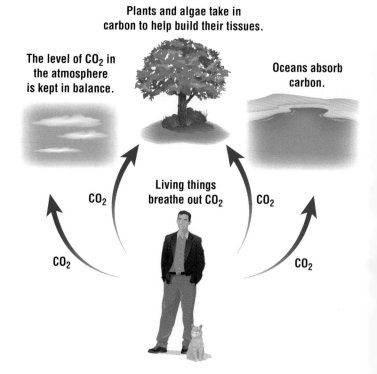

Plants and algae take in carbon to help build their tissues.

The level of CO_2 in the atmosphere is kept in balance.

Oceans absorb carbon.

Living things breathe out CO_2

CO_2

CO_2

CO_2

CO_2

balance. But he knew that burning wood, coal, and oil put carbon dioxide into the atmosphere faster than would happen naturally. This upset nature's balance. He predicted that if the carbon dioxide level doubled, the Earth's atmospheric temperature would increase by 41°F.

What's the controversy?

Though most scientists believe the Earth is warming up, not everyone agrees on the cause. Some scientists believe it is part of a natural cycle, while others think humans are to blame. Some believe that global warming will cause big changes in the weather patterns, while others think the changes will be small. And while most scientists believe global warming will be harmful, a few believe it will be helpful.

Why is there so much controversy? The atmosphere and the oceans form a complex system that affects global and local climates. The system is so complicated that it is difficult for scientists to make predictions. They use computers to model parts of the system according to their best data. But coming up with a good model is very hard.

Politics are also part of the controversy. Most scientists believe that humans are putting too much carbon dioxide into the atmosphere. To reduce carbon dioxide levels, scientists say we should reduce the amount of oil and coal we use. But those companies and nations that supply coal and oil do not want people to stop using these fuels.

Despite the controversy, most people today agree that the Earth is warming up. In this book you will learn how scientists study the Earth's atmosphere and climate and possible causes and results of global warming.

Chapter Two

What Causes the Greenhouse Effect?

The Earth's oceans and atmosphere are warmed mostly from **radiant energy**, which is energy from the Sun. It travels through space. When it reaches the Earth, you feel the Sun's energy as heat. You experience the effects of changes in the Sun's heat each year as the seasons change.

What are infrared waves?

About 60 percent of the radiant energy from the Sun is in the form of **infrared** radiation. Infrared is a type of radiant energy that humans cannot see but can feel as heat.

Most materials can absorb, or soak up, infrared radiation. This causes the objects to heat up and give off their own infrared radiation. The air around the heated objects then grows warmer.

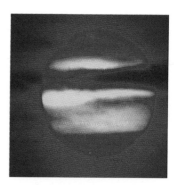

Looking at the Sun and clouds through an infrared lens, the red areas are the warmest, and the purple areas are the coolest.

Infrared waves can be reflected, or bounced, off surfaces. Because of this, infrared waves given off by the Earth's heated surface may be reflected back to the Earth by greenhouse gases in the atmosphere.

How does carbon dioxide trap heat?

Have you ever gotten into a car that has been sitting out in the Sun? You've probably noticed that the car is hot inside. Radiant energy enters the car's windows. The seats, dashboard, and other interior parts absorb some of the infrared radiation and heat up. Heat radiating from these solid parts warms the air. Some heat escapes through the windows, but much of the heat is reflected back from the windows and stays inside the car.

Carbon dioxide and other greenhouse gases are like a car's windows on a hot day. The Sun's rays pass through the atmosphere and warm the surface of the Earth. Heat, as infrared waves, radiates away from the warmed surfaces. Some of that radiated heat passes back through the atmosphere and is lost, while some of it is trapped by greenhouse gases. The more greenhouse gases in the atmosphere, the more heat that is trapped, and the warmer the entire Earth becomes over long periods of time.

The greenhouse effect

Venus

The planet Venus experiences an extreme version of the greenhouse effect. Venus has far more carbon dioxide in its atmosphere than the Earth does. The atmosphere of Venus is 96.5 percent carbon dioxide. Only 0.035 percent of Earth's atmosphere is carbon dioxide. All of this carbon dioxide causes a lot of heat to be trapped. While Earth's average temperature is 59°F, the average temperature of Venus is a scorching 867°F.

Are there other greenhouse gases?

Carbon dioxide is the main greenhouse gas. Other gases, however, help create the greenhouse effect. Though these gases are present in small amounts, they have a big effect when added together.

Methane

Methane is found naturally in the Earth, trapped in pockets in rocks. Methane is also a waste product given off by certain kinds

Brrr!
Without the greenhouse gases, heat would escape back into space. Earth's average temperature then would be about 60 degrees colder!

of bacteria, very simple organisms that can be responsible for diseases. Some of these bacteria live in the intestines of grazing animals, such as cows. Others live in wet soils, such as the mud in rice fields. Methane is also found in **permafrost** in the Arctic regions and in frozen deposits in the Arctic seafloor. Scientists have noticed more permafrost thawing in some Arctic regions and have seen large chunks of frozen methane hydrate floating to the surface of Arctic seas. If global temperatures increase, the Arctic may emit even more methane into the atmosphere.

What is methane hydrate?

Methane hydrate is a molecule of methane gas surrounded by a cage of water molecules. It looks like ice crystals.

Nitrous oxide

Nitrous oxide is created by the combination of nitrogen and oxygen. About one-third of the nitrous oxide in the atmosphere today is due to human activity. Much of it comes from the burning of coal, oil, and natural gas. The production and use of chemical fertilizers that help plants grow and the process of manufacturing nylon used to make brush bristles and clothing also produce nitrous oxide.

Uses for nitrous oxide

Nitrous oxide has been used as a mild anesthetic, a substance that lessens pain sensitivity, since the 1800s and has become known as "laughing gas." In the last half of the 20th century it also began being used to enhance performance of automobiles, especially for racing.

Ozone

Ozone on the ground is a **pollutant**. But ozone high in the atmosphere screens out ultraviolet rays. These high-energy rays in sunlight are damaging to living things. So ozone has both harmful and helpful qualities.

Chlorofluorocarbons

Chlorofluorocarbons, or CFCs, do not occur naturally. They are made by humans for use in air conditioners and to clean computer microchips. CFCs damage the ozone layer of the atmosphere, which is another global problem.

The ozone layer

The ozone layer of the atmosphere is found between 12 and 30 miles above the Earth's surface. In the 1970s, scientists discovered that CFCs could be damaging the ozone layer. After this, aerosol cans that used CFCs were banned. A "hole," or thinning of the ozone layer, has been reported in the Antarctic and Arctic regions.

Activity

Greenhouse effect in a jar

You can observe the greenhouse effect in miniature. You will need a large glass jar with a lid and two thermometers. Place one thermometer inside the jar, and put the lid on the jar. Leave the other thermometer outside the jar. Rest the thermometer outside the jar on a book so that the bulb doesn't touch the surface of the table or counter. (You want to measure the air temperature, not the temperature of the surface.) Set the jar and the outside thermometer in bright sunshine or in front of a very bright lamp. Check the temperature inside and outside the jar every 5 minutes for 30 minutes. Graph your results to compare them. Which was warmer—the air inside the jar or the air outside?

Chapter Three

Carbon Dioxide and Fossil Fuels

Scientists who study global warming are especially interested in measuring the amount of carbon dioxide in the atmosphere. Most scientists agree that the more carbon dioxide in the atmosphere, the warmer the planet.

How is carbon dioxide measured today?

Scientists at the Mauna Loa observatory in Hawaii have been sampling the Earth's atmosphere for over 30 years. Why Mauna Loa? The observatory sits 11,000 feet above sea level. Hawaii is in the middle of the Pacific Ocean, far from any continent. The air there is clear, with very little pollution.

Hawaii

Mauna Loa

On top of the observatory building is a 130-foot tower. The tower holds thin plastic tubes that collect air samples. Pumps inside the building draw air through the tubes. The air is collected in a freezer trap that cools the air to -174°F. This freezes the **water vapor** in the air but not the carbon dioxide. The remaining gas is pumped into a test chamber. Inside the test chamber, heating coils give off infrared radiation. This causes the air to heat up. The more the air heats up, the more carbon dioxide is in the sample.

13

The Mauna Loa Solar Observatory

Mauna Loa isn't the only place on Earth where the atmosphere is being tested. Scientists in Antarctica also sample the atmosphere and measure the amount of carbon dioxide. Because the research stations in Antarctica are far from heavily populated areas, the air is relatively pure, just as it is on Mauna Loa, Hawaii.

How was carbon dioxide measured in the past?

Scientists have come up with many creative ways of sampling the atmosphere from long ago. They have collected air from old bottles that were sealed on a known date. They have found hollow metal buttons on Civil War uniforms that were sealed so tightly that the air inside was preserved. Any sealed container from the past may have a useful air sample inside, which can help scientists determine differences in the atmosphere from past to present.

To go even further into the past, scientists use air samples from Antarctic and Greenlandic ice. In some areas on the Antarctic continent,

A scientist holds a glacial core sample, from which he will measure the carbon dioxide level.

the ice is two miles thick. It has been accumulating for thousands of years. Scientists use hollow drills that draw out long cores of ice. They sample bubbles of air trapped in the ice to find out what the composition of the atmosphere was when the bubbles were made. Then they use a scientific method called *Carbon-14 dating* to find out how old the ice is.

In 1999, a sample drilled out of a sheet of ice from Antarctica showed that greenhouse gas levels are higher now than at any time in the past 420,000 years.

How does Carbon-14 dating work?

Carbon-14 dating uses carbon that is radioactive to determine the age of fossils or other biological materials. Every living organism contains Carbon-14. When the organism dies, the Carbon-14 decays. Scientists can figure out how old a once-living organism is by determining how much the Carbon-14 has decayed.

What are fossil fuels?

The Carboniferous period occurred between 350 million and 290 million years ago. Where do you think the name *Carboniferous* comes from? If you guessed "carbon," you're right. Here's why.

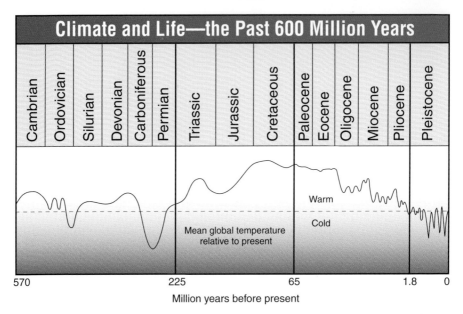

The period began with a very warm climate. The climate was also wetter than it is today. This made it perfect for huge forests and swamps to thrive. Plants take in carbon dioxide, just as humans take in oxygen. So the abundant plants took large amounts of carbon dioxide from the air. The carbon dioxide was then stored in the tissues of the plants.

When the plants during this period died, many did not decompose, or break down, completely. Much of the ground at this time was very wet. Decomposition did not totally occur in this wet environment because not enough oxygen reached the dead organisms. Oxygen is necessary for the decaying process. Dead plants sank to the bottom of swamps and oceans. These layers of dead plants and organisms became **peat**. After millions of years, some of this peat turned into coal, oil, and natural gas. They are known as **fossil fuels** because they were formed millions of years ago, even before dinosaurs roamed the Earth.

How are fossil fuels turned into energy?

It wasn't until a few thousand years ago that people realized that these fossil fuels could be turned into energy. Even so, burning fossil fuels didn't become a common practice until the 1800s.

The Industrial Revolution began in the early 1800s. This was not a military uprising but more of an economic and social change. It occurred in Europe and parts of the United States. Before this time, many people depended on agriculture as a way of life. But during the Industrial Revolution, people began manufacturing food and products. People moved from rural to urban areas. Many factories were built and provided new jobs. These factories needed energy to run. This is where fossil fuels came into play.

Factories burned coal to heat steam engines that ran the machinery. Later, gasoline engines and oil-burning furnaces were invented. Burning coal, oil, and gasoline adds carbon dioxide to the atmosphere.

Air samples from the time of the Industrial Revolution through today show that the level of carbon dioxide has increased. In 1850, the concentration of carbon dioxide was 288 parts per million. In 1990, it was 354 parts per million. While it is not absolutely certain that humans caused this change, human activity is one likely explanation. What is certain is that carbon dioxide levels are higher now than they were in the past.

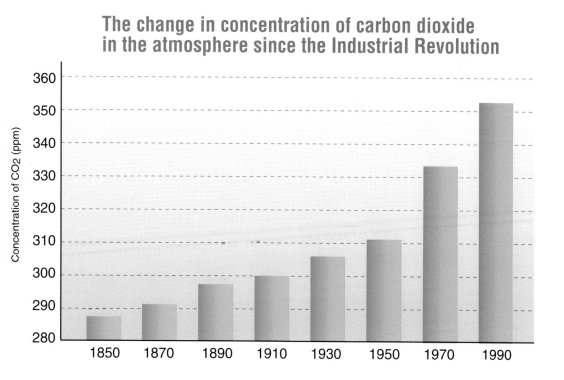

Chapter Four

The Changing Climate

In addition to studying carbon dioxide levels in the atmosphere, scientists who study global warming also look for evidence of global climate change. They compare the climate in the past to the climate today. Since the middle of the 1800s, people have kept accurate weather records all over the world. Discovering what the climate was like before then requires some real detective work.

How can we learn about the climate in the past?

Checking out fossils

Fossils, especially fossils of pollen, reveal clues about the global climate of the ancient past. Pollen is the powdery part of a plant used to reproduce, or make more, plants. By identifying pollen that has fossilized, scientists can tell what kinds of plants grew in an area in the past. However, fossils of pollen only allow scientists to estimate major changes in climate.

The changing of the seasons

Even before reliable thermometers were widely used, people were recording the climate in different ways. Weather was very important to people whose lives were tied closely to agriculture. Over many centuries, people have recorded the onset of spring and winter. They have also recorded harvest dates to tell when their crops were picked, and the dates that fruit trees and other crop plants bloomed. These written records tell us how far north of the equator people were able to grow various crops.

Clues of glacial proportions

How do scientists know when glaciers that are no longer there changed? As glaciers advance or grow, they leave behind certain scratches on the rocks beneath them. They also leave moraines, which are crescent-shaped hills of rubble that glaciers push in front of them. These clues help scientists tell when an area warmed or cooled.

Evidence from glaciers

Paintings and old photographs of mountains and their **glaciers** have also provided important information to climate researchers. Glaciers form in high mountain areas where the air is cold. Glaciers can only be found above the equilibrium snowline, which is the boundary at which snow melts as fast as it accumulates.

As snow builds up on top of the glaciers, it puts pressure on the ice below. This forces long tongues of the glacier to extend down valleys, well below the equilibrium snowline. These glacial tongues are sensitive to changes in climate. They may advance or retreat hundreds of feet in response to small changes in climate. For example, climatologists know from old photographs that the Rhône glacier in the Alps has retreated three miles since 1860. Global temperatures since that time have changed only 1 or 2 degrees. On the remote Arctic island of Svalbard, 375 miles north of Norway, a glacier that was photographed in 1918 has almost disappeared completely.

Rhône glacier

Tree rings

One way scientists learn about climate changes in the recent past is by looking at tree rings. You probably know that you can tell how old a tree is by counting its rings. But did you know that you can also tell how much precipitation fell each year? Precipitation is moisture that falls to the ground, as in rain or snow. By studying the size of the tree rings, scientists can determine changes in precipitation and temperature. Large rings indicate good weather and precipitation.

What do you want to be when you grow up?

A scientist who specializes in reading tree rings is called a *dendrochronologist*.

How do we learn about the climate today?

Scientists can tell what is currently happening in the atmosphere from weather stations, weather satellites, ocean buoys, and weather balloons. Weather stations are set up on land. They can measure temperature, wind, and precipitation. Weather satellites can also measure these things, but they measure them from space. They send the information, along with pictures, back to scientists on Earth. Ocean buoys float on the water. They give scientists information about air temperature, water temperature, and conditions of the atmosphere.

Weather balloons are useful in studying what's happening in the atmosphere. Different types of weather balloons are used to measure different conditions, such as wind speed and direction, clouds, atmospheric pressure, temperature, and humidity. Some weather balloons release a parachute when they reach a certain altitude. The parachute measures weather conditions as it falls to the ground. Smaller balloons can only carry up to a pound of measuring equipment, but larger ones can carry up to 30 pounds.

Weather balloon

So what have we learned?

One thing scientists have learned from these kinds of evidence is that the Earth was a little warmer than it is today during a period they call the Medieval Warming Period. During this time, the Vikings explored the North Atlantic. They established a colony in Greenland, where they were able to grow crops and graze sheep.

Around 1300, however, the Greenland colony failed. One reason for this was that the weather turned cold, making it hard for the Vikings to grow food. This was the end of the Medieval Warming Period and the beginning of a cooling period that lasted until about the middle of the 1800s. The period is known as the Little Ice Age, not because it was short compared to other ice ages, but because it involved only a small change in global temperatures. That small change, however, had big effects on people. Harvests were smaller and crop failures occurred in Northern Europe. Famines, when hunger was widespread, reduced the population and caused political changes.

> **Just exactly who were the Vikings?**
>
> Vikings were Scandinavian people who not only explored areas by ship but also raided and invaded parts of northwestern Europe. Vikings lived from the 8th to 11th centuries A.D.

Activity

Looking at historical changes

What evidence of climate change in your community does history reveal? Find historical pictures and written records from your community that date back over the last century or longer. Your local library or historical society may help you find photos. Art museums may have old paintings of your area. Your library may also have old newspapers on microfilm. Using these historical documents, what evidence can you find of changes in climate in your local area over time? Clues you can look for include pictures of winter snowfall, weather records in old newpapers, and the kinds of clothing people wore in old pictures. The dates of historical photos can help you also. What time of year did harvests occur? How early did winter start or how late did it last?

Chapter Five

The Effects of Changing Temperatures

Scientists have plenty of evidence that global warming is a reality. Scientists are also reasonably certain that increases in the Earth's temperatures are related to carbon dioxide levels in the Earth's atmosphere. But there are many other factors that affect global temperatures. This makes the study of global warming very complex, which causes disagreements over issues surrounding global warming.

Is the greenhouse effect the cause of global climate change?

Scientists have shown that both global temperatures and the level of carbon dioxide have increased over the past 150 years. They have also shown that higher levels of carbon dioxide in the atmosphere occur at the same time as higher global temperatures. But that doesn't mean that one causes the other.

In fact, no one knows for sure if the rise in temperature is due to the increased levels of carbon dioxide. Some scientists believe that the warming is just part of a natural cycle of warming and cooling. They cite other periods in history when temperature changes occurred.

One thing we can be sure of is that since 1850 humans have added large quantities of greenhouse gases to the atmosphere. Also, the addition of these gases has occurred at a much higher rate than would have happened naturally. Scientists are also confident that global temperatures have increased since 1850. What remains controversial is whether the rise in temperature is due to human activity.

Will global warming have positive or negative effects?

Just as scientists disagree about the cause of global warming, they also disagree about the effects it will have. While most believe that global warming will have negative effects on the Earth, some think that global warming is not necessarily a cause for alarm.

Those who feel that the effects could be positive point out that human culture flourished during the Medieval Warming Period. Agricultural production was high in Northern Europe. An earlier warming period is associated with the rise of agriculture and civilization in the Middle East and the transition from the Stone Age to the Bronze Age. These scientists suggest that today's agriculture would benefit from global warming. Higher overall temperatures would increase the growing season in most areas. Increased carbon dioxide in the atmosphere would be good for plants.

Ages that were ages ago

The Stone Age is the first known period of human culture. It occurred during prehistoric times and was when stone tools were first used. The Bronze Age is the first period in human culture in which metal was used. It occurred in Greece and China before 3000 B.C. but not until 1900 B.C. in Britain.

However, this view is not held by most scientists. Scientists all over the world warn that rising temperatures do not necessarily mean that we will have more pleasant summer days to enjoy.

What about oceans and global warming?

Coastal flooding is increasing on island nations such as the Marshall Islands and Tuvalu. Rising ocean water is also threatening crops on these islands as seawater contaminates the groundwater, making it unclean or harmful.

Why is this happening? There are two reasons. First, as ice on landmasses such as Antarctica melts, it causes sea levels to rise. Second, the greenhouse effect is causing oceans to become warmer. As water becomes warmer, it expands. This is because the water **molecules** move faster, causing them to spread farther apart. So warmer water takes up more space.

So what if sea levels rise? Nearly half of the world's people live on coasts. In fact, 15 of the world's 20 largest cities are in coastal regions. Rising seawaters would threaten these cities and the people who live there. But what about small islands? Some island nations have already begun trying to convince the world about the importance of reducing the greenhouse effect. Islands such as the Bahamas, Cuba, and Jamaica realize they risk being wiped out completely.

25

Meet the Inuits

Inuits are people who live on the Atlantic coast of Canada and Greenland and whose ancestors lived in the area before European settlers arrived.

Melting sea ice is also causing problems. Inuit hunters in the Arctic have noticed lately that the sea ice is thinner and melts earlier in the season. The Inuits hunt on the sea ice, so melting ice limits their hunting season.

What about weather and global warming?

Changing temperatures will also mean changes in weather. A warmer world could be a world with more extreme weather.

Warmer temperatures cause more water to **evaporate** into the air. In turn, this causes more precipitation. However, this doesn't mean that every place in the world will get more rainfall. The shifts in weather patterns may even cause drought in some areas. Indonesia is normally a very wet country, but in the 1990s it experienced a drought. This led to massive fires in the country's tropical forests.

Warmer temperatures also lead to increased violent weather. This has already begun to happen. In the 1900s, there were an unusually high number of tornadoes, hurricanes, and floods.

Activity

Polar ice in a pan

Is there a difference between melting sea ice and melting land ice? Does melting sea ice really cause flooding? After all, when ice in a drink melts, it does not change the volume of the drink.

You can find out if there is a difference between sea ice and land ice by doing a simple experiment. Pour water into two empty cottage cheese cartons to a depth of about 1 ½ inches. Freeze both until the ice is solid. Fill an 8"x8"x2" baking pan ¾ full of water. Add one chunk of ice. This represents floating sea ice. Measure the water level with a ruler and write it down. Let the ice melt completely, which will take several hours. Measure the water level again. Did it change?

Next you will model Antarctic land ice. Empty the baking pan. Press modeling clay into the empty cottage cheese container to make a "continent" that is about 2 inches thick. Stick the "continent" to the bottom of the baking pan. Add water to the pan as before. Put the second ice chunk on top of the "continent." Measure the level of water in the pan before and after the continental ice melts. What happened to the water level this time?

What about diseases and global warming?

Health care workers in England and Wales are concerned over cases of malaria that have been reported in those countries. Malaria is usually associated with tropical countries. However, as the global temperature rises, the mosquitoes that carry malaria can survive farther north of the equator than before. Warmer temperatures also mean the parasite that causes malaria matures faster.

In the United States, the West Nile Virus appeared for the first time in 1999 and spread rapidly during the summer of 2002. The mosquitoes that carry the virus are able to survive in many areas of the United States. It is possible that rising temperatures could allow other tropical diseases to spread.

A disease you don't want to get

Malaria is a serious disease that is spread when mosquitoes carrying the disease bite a human. The disease causes chills, shaking, and an intense fever.

But could temperatures drop?

Some scientists who believe the recent warming is due to natural causes blame the thermohaline circulation theory. Thermohaline circulation drives currents in the world's oceans. *Thermo* means "heat," and *haline* means "salt." Ocean water that is colder or saltier tends to sink. Cold ocean water sinking in the North Atlantic affects ocean currents all over the world. As the cold water sinks, it draws the warm Gulf Stream northward. Many scientists believe that during the Medieval Warming Period, the Gulf Stream was farther north than it is today. This made Greenland and Northern Europe warmer.

However, as global temperatures increase, ice in Antarctica melts at a higher rate. The rivers of cold water flowing off Antarctica cause ocean water in the southern Pacific and Atlantic to sink.

What is the Gulf Stream?

The Gulf Stream is a warm current of the Atlantic Ocean. It starts in the Gulf of Mexico and flows northeast along the coast of North America and beyond.

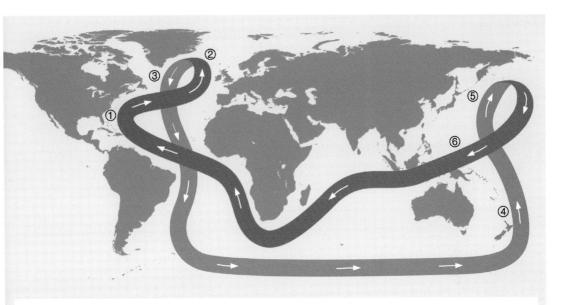

① The Gulf Stream pushes warm, salty surface waters north. ② The waters lose heat to the atmosphere. The surface waters cool and sink all the way to the bottom of the ocean. ③ The deep, cold waters flow southward to the Southern Atlantic. ④ When the waters get to the bottom of the Pacific Ocean, they drift north. ⑤ The cold bottom waters warm and rise to the surface. ⑥ The warmer waters flow back into the Atlantic Ocean, where they eventually meet the Gulf Stream again. This round trip can take up to 1000 years!

If the theory about thermohaline currents is correct, rising global temperatures may reverse ocean currents all over the world. One result could be that the Gulf Stream would drop southward. This may explain why Northern Europe and Greenland became cold and the Little Ice Age began. Scientists know that Antarctic ice is melting at higher rates today than it was a century ago. Some scientists believe this means global temperatures could actually *drop* in the next few decades. Instead of a warmer world, the world could actually experience another Little Ice Age!

Ice ages can make life difficult for humans. Cold northern climates reduce the amount of food that can be grown. Glaciers creeping across continents and down mountain valleys cool the surrounding areas and mow down forests.

Extreme climate changes, however, have inspired humans to invent new ways to survive. Humans have adapted by making new inventions and changing their behaviors. But can other life forms do the same?

Can plants and animals adapt?

The word *adaptation* is a confusing one, as the word has so many meanings. In everyday life, we use *adapt* to mean "to get used to something." When scientists talk about plants and animals "adapting," they are talking about the process of natural selection. Animals in the wild cannot simply decide to change their ways. If the climate turns cold, for instance, animals cannot decide to build shelters or grow thicker fur.

Adaptation is a slow process. Natural adaptation doesn't happen to individuals. It happens to populations over many generations. If the climate turns cold, mammals may migrate, or move to a new environment. If they cannot migrate, those mammals with thicker fur will be more likely to survive. Their offspring may inherit their thick fur. Those offspring in turn will have better odds of survival and be more likely to reproduce. It takes many generations for a population of animals to adapt to changes in their climate.

If the climate changes too fast, animals and plants that cannot migrate to new areas may die out. This is why rapid climate changes result in extinctions.

The temperature on the Antarctic Peninsula has risen over 36°F over the past 50 years. Animals, such as penguins, cannot adapt this quickly. The numbers of Adélie penguins have fallen by 19 perent in the past 25 years.

Polar bears in the Arctic are a good example of animals threatened by rapid climate changes. Polar bears live in the extreme north on the Arctic ice. As the ice melts, the polar bears' habitat is reduced. They cannot find enough food, which makes it hard for them to survive and to reproduce. Polar bears are adapted to the far north. They cannot simply decide to live somewhere else or eat something else. If the climate changes too rapidly, polar bears could become extinct.

Polar bear, polar bear, where are you?

Follow two female polar bears, Lena and Yana, on the Polar Bear Tracker maintained by the World Wildlife Federation (WWF). You can get a detailed report of where and how far each bear has traveled, along with maps of the Arctic area. You can also find out information about global warming and how it is affecting the polar bears' habitat. Go to www.panda.org/polarbears/.

Lean times for polar bears

Research in the Hudson and James Bay areas in Canada shows that the amount of time polar bears have on the ice is getting shorter. Spring is coming earlier, and autumn is starting later. For every week earlier that the ice melts, bears come to shore an average of 22 pounds lighter and in poorer health.

Chapter Six

What Does the Future Hold?

Scientists have plenty of evidence that global warming is a reality, and concerned people are choosing not to just sit back and let global warming happen. Many scientists, businesses, politicians, and individual citizens who feel greenhouse gases are the cause are working to reduce them and decrease the greenhouse effect.

What are scientists doing?

Besides measuring the greenhouse effect, science can help us find ways to reduce it. Climate researchers are developing new, more accurate models of the global climate to help everyone make better decisions. Industrial engineers are developing new ways to reduce **emissions** from factories. One research team in Australia is even developing a way to put waste carbon dioxide into the Earth.

Promoting alternative power

Scientists continue to research alternative forms of energy to avoid using fossil fuels. Wind power is practical in windy areas,

Wind power

Solar power

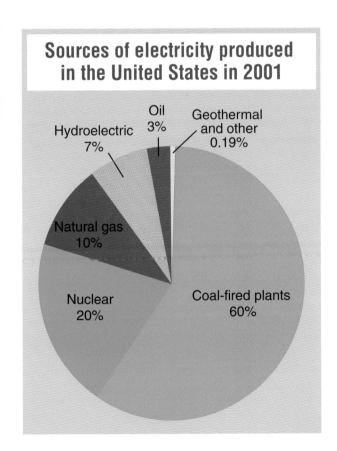

Sources of electricity produced in the United States in 2001

- Oil 3%
- Geothermal and other 0.19%
- Hydroelectric 7%
- Natural gas 10%
- Nuclear 20%
- Coal-fired plants 60%

such as coastal areas and the Great Plains. Solar energy could generate electricity in sunny areas of the country and turn rooftops into energy generators.

Hydrogen fuel cells are a promising new line of research. Hydrogen is a gas. It is colorless and odorless. It is very abundant in the universe, but it is only found in combination with other elements. So to use hydrogen, scientists must first separate it from those other elements.

Hydrogen burns cleanly and gives off water vapor, not carbon dioxide. This is why the development of hydrogen as a fuel is so exciting. When it burns, it doesn't damage the environment. Hydrogen fuel cells are expensive but produce a lot of energy. They convert hydrogen into electricity. This technology is being used to develop new, more energy-efficient automobiles. Someday hydrogen fuel cells could power aircrafts and our homes.

Energy that's out of this world

Hydrogen is used by NASA as fuel for the space shuttles. Hydrogen fuel cells that provide heat, electricity, and drinking water for the astronauts are also being used by the space program.

The Daimler Chrysler NECAR is powered by hydrogen.

Studying plants and carbon dioxide

Some scientists are researching how plants absorb carbon dioxide from the atmosphere. Many people who are worried about the environment plant trees to help reduce greenhouse gases. But scientists are finding that trees aren't always the best choice. Some are finding that replacing grass with trees in areas where prairies grow naturally can actually decrease the amount of carbon that is stored. Many people don't realize how much of a grass plant is below ground. The extensive root system requires a lot of carbon to build. Also, native grasses are adapted to prairie areas. They thrive, while non-native trees require care.

To study the effects of increased carbon dioxide, researchers exposed these trees to double levels of carbon dioxide. This is the level of carbon dioxide in the atmosphere expected a century from now.

What are corporations doing?

Some corporations are doing very little to change their ways. Reducing greenhouse gas emissions can be costly, and some corporations don't want to increase their operating costs. They work hard to convince politicians not to enact laws to reduce greenhouse emissions.

Other corporations are trying to change. They have leaders who believe in doing their part to take care of the environment. Other corporate leaders believe that if they are working hard to reduce the greenhouse effect, people will want to buy their products and support them.

Some companies develop products that reduce greenhouse gases and help people save money at the same time. Automobile manufacturers are making cars more energy-efficient. Fuel-efficient cars produce less carbon dioxide and save money in the long run. Appliance manufacturers are making appliances more energy-efficient than ever. Light bulb manufacturers are making fluorescent light bulbs that fit into regular light bulb sockets. These bulbs use less energy, so even though they are more expensive to buy, they pay for themselves in energy savings. Companies are also developing lights using Light Emitting Diodes (LEDs) instead of bulbs. LEDs use very little electricity but put out a lot of light. New flashlights and holiday lights using LEDs have already hit the market.

What are politicians doing?

Politicians from many nations are working together to try to reduce greenhouse gases.

Kyoto Protocol

One effort toward cooperation is an international treaty called the Kyoto Protocol. Participating countries would reduce their greenhouse gas emissions from 1990 levels by the year 2012. After the initial signing of the agreement, countries began disagreeing about how the treaty would be put into place. After talks broke down, the United States, under George W. Bush, pulled out of the treaty. However, many industrialized nations, such as Japan, the Czech Republic, and Russia, have agreed to the treaty.

> **What's Kyoto?**
>
> This legally binding treaty is called the Kyoto Protocol because leaders from 150 countries signed it in 1997 in Kyoto, Japan.

Brazil and Germany are two countries cooperating to meet the Kyoto goals. In 2002, Germany agreed to supply Brazil with cars that burn alcohol as fuel. One of Brazil's major industries is sugar. After sugar-rich juices are squeezed from sugar cane, the remaining waste is usually burned. The Brazilian government decided the country could reduce its dependency on foreign oil and reduce carbon dioxide emissions if the sugar cane waste was used to make alcohol. The alcohol can then be used to fuel cars.

U.S. efforts

The United States decided not to follow the Kyoto Protocol. However, the United States is committed to addressing the issue of global warming.

President George W. Bush developed a plan for dealing with global warming. His plan is designed to reduce greenhouse gas emissions while protecting the growth of the economy. A growing economy means new technologies to replace ones that are hurting our environment.

Japanese Prime Minister Junichiro Koizumi met with British Deputy Prime Minister John Prescott about the Kyoto Protocol.

What can you do?

If you think global warming is a problem, there are things you can do.

The United States alone uses ¼ of the world's energy, most of which comes from fossil fuels. When we use up these fossil fuels, we can't use them again. They were created millions of years ago and cannot be re-created. Plus, these fossil fuels release tons of carbon dioxide back into the environment.

The choices individuals make can have a surprisingly powerful effect. When people choose to buy more efficient cars and appliances, they send a message to big industries that efficiency is important.

Be a wise consumer

All the products you buy required energy to be produced. The more packaging a product has and the farther a product had to travel, the more energy was required to put it on the shelf.

That's a lot of hot air!

On the average, each American releases 20 tons of carbon dioxide into the air each year.

Fuel-Efficient Cars	MPG City	MPG Hwy
Most Efficient Overall Honda Insight	61	68
Most Efficient Compact Car Toyota Prius	52	54
Most Efficient Small Station Wagon Volkswagen Jetta Wagon (Diesel)	42	50
Least Efficient Compact Car Bentley Continental R	11	16
Least Efficient Small Station Wagon Bentley Continental R	17	21

Source: U.S. Department of Energy

You send greenhouse gases into the air when you:

watch TV
use the air conditioner
turn on a light
use a hair dryer
ride in a car
play a video game
listen to a stereo
wash or dry clothes
use a dishwasher
microwave a meal

Source: U.S. Environmental Protection Agency

Buying products that use less packaging and products made locally helps save energy. You can also write to manufacturers and ask them to reduce their packaging.

Buying fewer products in the first place also reduces energy use. Watch advertisements on television. How many try to convince you to buy products no one really needs? Advertisers have created an inflated sense of our needs. Shopping is a popular pastime in this country, and Americans buy far more products than people in other nations.

Of course, buying fewer products saves *money* too. Conserving energy can also save money. The average American family spends $1300 each year on utility bills for their home. And much of this energy is wasted.

Make efforts to conserve energy

So how can you conserve energy? There are many ways. Some are very simple, and others require a bit more planning. Some you can do yourself, and others may require educating the adults in your life at home and at school. But all contribute to a healthier planet!

- Turn things off! Turn off lights when you're not using them. The same goes for computers, stereos, and video game systems. Even if you leave a unit on "standby," it still uses half the energy it does when it's running.

- Have your parents turn down the thermostat in your home. Also, make sure no heat or cold air is escaping through gaps in windows or doors. Insulating your home saves wasted energy and money!

- Ask your parents to purchase energy-efficient light bulbs. The standard light bulbs that we use today are not very different from the one Thomas Edison invented in 1879. In fact, only 10 percent of the energy used by one of these bulbs is used for making light. The other 90 percent of the energy is wasted as heat. One type of

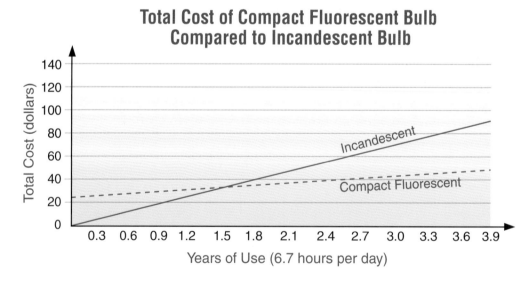

The incandescent bulb is much cheaper to purchase. However, it doesn't last as long as a compact fluorescent bulb. It also isn't as energy-efficient. This makes the long-term cost of the compact fluorescent bulb less.

energy-efficient light, called *compact fluorescent light bulbs*, uses 66 percent less energy than standard bulbs. It also lasts up to 10 times longer! This helps the environment and the economy of your household too!

- Recycle! Does your family recycle its newspapers, cardboard, glass, and metal? By doing so, you can reduce the amount of carbon dioxide that goes into the air by 850 pounds each year! And when you're buying new products, don't forget to look for ones that use reusable, recyclable, or reduced packaging. This saves the energy required to manufacture new containers.

- Consider walking or riding a bike when you can instead of having your parents drive you places. Encourage your parents to walk to work or carpool. Every gallon of gasoline used releases 22 pounds of carbon dioxide into the air.

- Planting native plants around your house helps reduce carbon in the atmosphere. Deciduous trees, ones that lose their leaves in the fall, shade your house in the summer and let sun in during the winter, helping you save energy. Native plants are also good for local wildlife.

- Have you ever grown a vegetable garden? Or bought produce from a local fruit stand? Some fruits and vegetables you see at the grocery store are imported from countries far away, requiring large amounts of fuel to get them to you. Save energy by growing some of your own food or buying from local farmers.

Activity
Hidden impacts

List some of the products you like to buy frequently, such as candy, CDs, magazines, or clothing. Find out who makes these products. Look up the companies on the Internet. Search their Web sites to see if they have an environmental policy. If you can't find any mention of such a policy, email the companies to find out what they are doing to reduce greenhouse emissions. Many company Web sites have feedback forms you can use. If you find out that a company that makes one of your favorite products isn't interested in reducing emissions, will that affect your buying decisions?

Conclusion

Global warming continues to be a hotly debated issue. While it's readily accepted that the Earth is getting warmer, scientists continue to disagree about what that really means. Are we causing it, or is it just happening naturally? Is it going to be harmful to our planet, or could it be helpful? What's in store for us if the Earth's temperature continues to rise? Read the evidence on both sides and decide for yourself.

Internet Connections and Related Readings

Climate Kaleidoscope and the Rain Forest (http://climate.nms.ac.uk/)

Get your questions answered about global warming and specifically how trees and rain forests affect it. Meet the scientists who are studying the rain forest and find out what they do, how they do it, and why they do it. See pictures of the rain forest and keep up on the latest news and developments in global warming and the rain forest.

Cool Climate Kids' Club (http://www.coolclimate.org)

Inspired by two local sixth graders who were worried about climate change, this Ontario-based Web site is anything but boring! Explore the subject of global warming in "What's with this climate change?" Find out how you can help. You can also learn some wacky facts about climate change, read some jokes, solve a crossword puzzle, or sing a song. A glossary will give you all the definitions you need to be a global warming expert.

EcoKids Online (http://www.ecokidsonline.com)

Open the Eco-Info Notebook to browse environmental issues, including climate change. Carbon dioxide is taking over the world! Only you can stop it from warming up the world too much and changing the face of the planet forever by playing a cool interactive game! Print project sheets, coloring pages, and puzzles. This Web site is chock full of informative and entertaining activities about everything ecological.

**EPA Global Warming Kids Site
(http://www.epa.gov/globalwarming/kids/)**
This Web site explains the greenhouse effect, climate, and global warming, including how you can make a difference. You can also do a global warming word search, play Hangman against Ozone the Dog, solve a global warming crossword puzzle, or test your concentration with pictures related to global warming.

**Who's Professor Polar Bear?
(http://www.arm.gov/docs/education/whoisprof.html)**
Professor Polar Bear and his scientist friends will answer your questions on the warming environment and how it relates to melting ice. Whether you're a global beginner, a global thinker, or a global expert, this site has something for you. Ask a scientist a question, send a postcard to a friend, or take a quiz on what you've learned.

Ecology by Steve Pollock. Eyewitness Science. Dorling Kindersley, 1992. [RL 7.6, IL 3–8] (5868306 HB)

Green Thumb by Rob Thomas. Grady Jacobs, 13, is a nerd who's dying to change his life. So when an invitation comes to join Dr. Carter's team in the Brazilian rain forest, Grady is on the next plane to the Amazon. Grady quickly sees that Carter's experiments are destroying the forest, and he knows he's got to do something. Simon & Schuster, 1999. [RL 5.5, IL 5–9] (3188002 CC 3188001 PB)

The Greenhouse Effect by Alex Edmonds. Exciting visual images, anecdotal text, and hands-on projects enhance current research on the subject of the greenhouse effect and encourage students to unravel the issues and take responsibility for ecological safety. Includes maps, diagrams, glossary, and index. Millbrook Press, 1997. [RL 5, IL 4–6] (03109706 HB)

Glossary

atmosphere (AT muhs feer) the layer of gases that surrounds a planet, including the Earth

carbon dioxide (KAR buhn deye OK seyed) a **greenhouse gas** made up of carbon and oxygen (see separate entry)

chlorofluorocarbon (klor oh FLOR oh kar buhn) also known as CFC, a **greenhouse gas** made by humans for use as a refrigerant or in aerosol cans (see separate entry)

climate (KLEYE muht) the long-term average weather conditions

drought (drowt) a long period of unusually dry weather

emissions (ee MISH uhnz) harmful or impure substances, including greenhouse gases, given off by vehicles and factories

evaporate (ee VAP er ayt) to change from a liquid into a gas

fossil fuel (FAW suhl fyoul) coal, oil, or natural gas buried deep in the Earth as the result of the decomposition of prehistoric plants and animals

glacier (GLAY shur) a large, long-lasting, moving mass of ice made up of accumulated snow

greenhouse effect (GREEN hows uh FEKT) the heating of the atmosphere caused by **greenhouse gases**, including **carbon dioxide** (see separate entries)

greenhouse gas (GREEN hows gas) any of the gases that contribute to the **greenhouse effect** by trapping heat in the **atmosphere** (see separate entries)

hurricane (HER uh kayn) severe tropical storm

ice age (eyes ayj) any period in history when the Earth cooled and large glaciers formed on the continents

infrared (IN fruh red) a type of **radiant energy** that is felt as heat (see separate entry)

methane (METH ayn) a **greenhouse gas** found in rocks or given off by certain kids of bacteria (see separate entry)

molecule (MAHL uh kyoul) the smallest physical unit of a substance that can exist by itself

nitrous oxide (NEYE truhs AWKS seyed) a **greenhouse gas** whose **molecules** are made up of nitrogen and oxygen (see separate entries)

ozone (OH zohn) a **greenhouse gas** that is helpful high in the **atmosphere** but is a **pollutant** on the ground (see separate entries)

peat (peet) deposits of dead plants in the bottoms of swamps that build up over thousands of years

permafrost (PER muh frahst) rock or soil underneath the Earth's surface that remains permanently frozen

pollutant (puh LOO tuhnt) something that causes something else to be impure or unclean or is harmful to something else

radiant energy (RAY dee uhnt EN er jee) energy from the Sun

water vapor (WAH ter VAY per) water that is in the form of a gas

Index

adaptation, 29–31
alternative power
    hydrogen, 33–34
    solar energy, 33
    wind power, 32–33
Antarctica, 12, 14–15, 25, 27, 28, 30
Arctic, 11, 12, 26, 31
Arrhenius, Svante, 6–7
Bronze Age, 24
Bush, George W., 36, 37
Carbon-14 dating, 15
Carboniferous Period, 15–16
diseases
    malaria, 28
    West Nile Virus, 28
fossil fuels, 15–17, 32, 38
fossils, 18
Fourier, Joseph, 6
glaciers
    equilibrium snowline, 19–20
    Rhône glacier, 20
    Svalbard glacier, 20
greenhouse effect, 5–6, 8–12, 23–24, 25, 32, 36
greenhouse gases
    carbon dioxide, 6–7, 9–10, 13, 14–15, 17, 23–24, 35, 38, 41
    chlorofluorocarbons, 12
    methane, 10–11
    nitrous oxide, 11
    ozone, 12

Gulf Stream, 28–29
ice age, 5, 22, 29
Industrial Revolution, 16–17
infrared radiation, 8–9
Inuits, 26
Kyoto Protocol, 36–37
Little Ice Age, 22, 29
Mauna Loa, Hawaii, 13–14
Medieval Warming Period, 21–22, 24, 28
NASA, 4, 34
ocean buoys, 21
peat, 16
polar bears, 31
radiant energy, 8–9
sea levels
    coastal flooding, 25
    islands, 25
Stone Age, 24
thermohaline circulation, 28–29
tree rings, 20
Venus, 10
Vikings, 22
weather
    balloons, 21
    extremes, 26
    satellites, 21
    stations, 21